The Essential Oil Diffuser Recipes Book

The Essential Oil
Diffuser Recipes Book

*Over 200 Diffuser Recipes for
Health, Mood, and Home*

Julia Grady

Copyright © 2017 by Dylanna Publishing, Inc.
All rights reserved. This book or any portion thereof
may not be reproduced or used in any manner whatsoever without the express written permission of the publisher except for the use of brief quotations in a book review.

Julia Grady

First edition: 2017

Disclaimer/Limit of Liability

This book is for informational purposes only. The views expressed are those of the author alone, and should not be taken as expert, legal, or medical advice. The reader is responsible for his or her own actions.

Every attempt has been made to verify the accuracy of the information in this publication. However, neither the author nor the publisher assumes any responsibility for errors, omissions, or contrary interpretation of the material contained herein.

This book is not intended to provide medical advice. Please see your health care professional before embarking on any new diet or exercise program. The reader should regularly consult physician in matters relating to his/her health and particularly with respect to any symptoms that may require diagnosis or medical attention.

Contents

Introduction 9
Essential Oil Basics 10
Recipes 17
Energize 19
Sleep 22
Relaxation and Calmness 25
Mood Enhancement 28
Stress Relief 32
Colds and Congestion 35
Immune System Boosters 38
Headache Relief 41
Cognitive Function 44
Insect Repellant 47
Meditation 49
Appetite Suppressant/Fight Hunger Cravings 51
Air Fresheners 53
Beat the Heat 56
Positive Thinking 58
Anxiety 61
Depression 65
Indigestion and Nausea 68
Holiday/Winter Blends 71
Summertime Blends 74
Autumn Blends 77
Springtime Blends 80
Resources and References 83

Introduction

Essential oils have been used for millennia and are useful for a variety of purposes. These all-natural oils are an excellent complementary and alternative approach to improving health, mood, and the home and when used properly, they have very few side effects. As more people discover the many uses and benefits of using essential oils, they will continue to increase in popularity.

Aromatherapy is a method that uses the sense of smell to influence the brain and body to relieve pain and stress, boost mood, increase energy, fight fatigue, decrease appetite, lose weight, and a multitude of other purposes. Every essential oil has its own unique properties and produces different results. For example, eucalyptus may be used as a pick-me-up after a long day, while lavender will help you relax and fall asleep.

The sense of smell is very powerful. It can trigger memories and influence emotions. Smells travel by way of the olfactory nerves up to the limbic area of the brain. This part of the brain is responsible for our moods, feelings, and memory, as well as conditioned responses and learning. When the limbic system is stimulated, it releases neurotransmitters, endorphins, and other "feel-good" chemicals. In this way, aromatherapy and essential oils can alter the brain's chemistry.

Essential Oil Basics

What Are Essential Oils?

They are the natural oils that are extracted from plants, including from the leaves, stem, flowers, fruits, bark, and even roots. Do not confuse them with perfumes, which are fabricated substances without healing properties. Unlike perfumes, essential oils do not contain any synthetic ingredients and they have known benefits for both the mind and body.

Essential oils contain various chemical components that influence the different systems in the body. This may sound somewhat dangerous, but it actually isn't because our bodies naturally produce and use many of these same chemical compounds already. The essential oils work by stimulating or sedating the body's systems. Many also contain powerful anti-inflammatory, antiviral, antibacterial, antifungal, and other properties.

For example, ester, a chemical component found in bergamot, chamomile, lavender, and sage, acts as a sedative, a calming agent, an antifungal, and an anti-inflammatory. Ketones, substances that promote new cell growth and help wounds heal faster, are found in camphor, eucalyptus, rosemary, and sage. Alcohols, which act as diuretics and fight bacteria, are contained in ginger, patchouli, peppermint, rose, rosewood, sandalwood, and tea tree, as well as other essential oils.

Science of Aromatherapy

Aromatherapy is the practice of using the essences derived from beneficial plants to heal the body. It is also known as essential oil therapy because the aromatic essence is obtained from essential oils. The oils may be inhaled, applied topically, and in some cases ingested.

Aromatherapy is based on the use of plants to balance and harmonize the psychological, spiritual, and physical aspects of the human body. Once this balance is achieved, the body can better heal itself.

When an essential oil is inhaled its scent is carried through the nose to the brain via olfactory nerve cells. Upon reaching the brain, the scent activates the limbic center, which is the part of the brain responsible for emotions, among other functions. Neurotransmitters such as serotonin and dopamine, as well as endorphins, are released Depending upon the essential oil inhaled, you may feel aroused, excited, or more relaxed.

Benefits and Uses of Essential Oils and Aromatherapy

Aromatherapy is commonly used to promote general relaxation, increase feelings of well-being, and reduce stress. While aromatherapy is very effective for altering mood, essential oils can have positive effects on many other levels. They are also used to boost the immune system and fight inflammation throughout the body. In addition, they can be used to treat infections, relieve pain, and promote healing.

There are hundreds of different essential oils, many of which are only occasionally used and some of which are known poisons. There also a multitude of known uses. Here we list some of the most common uses of essential oils.

Mood Enhancement

One thing many essential oils are used for is the regulation of mood. Ranging from the ability to arouse to the ability relax, essential oils can lead to a wide variety of changes in how you perceive your environment.

Muscle Relaxation

Essential oils are frequently used in massage oils and lotions, making them particularly useful for treating muscle discomfort. They can be used to reduce muscle spasms, promote healing, and aid in overall relaxation. When combined with massage they are a powerful form of therapy.

Skincare

The antibacterial, anti-inflammatory, and antiseptic properties of many essential oils make them an ideal addition for skincare products. When added to cosmetics, balms, or lotions, certain oils can help prevent scars, combat acne, heal dry skin, reduce the appearance of stretch marks, ease rashes, and promote wound healing.

Pain Relief

Essential oils can also be a powerful source of pain relief. Many are natural analgesics, containing substances such as methyl salicylate the main ingredient in aspirin, and when used can block the body's pain receptors. Wintergreen oil contains a high concentration of methyl salicylate and it has been documented that this oil can have cortisone-like effects that can get to the root of pain and provide fast relief. Wintergreen has been used for centuries as an effective pain reliever.

Anti-Inflammation

Chronic inflammation is thought to be the root cause of many diseases today including heart disease, cancer, diabetes, and arthritis. Research has shown essential oils to have the ability to reduce inflammation.

COX enzymes are responsible for the body's inflammatory response. Several essential oils work as proven COX inhibitors, reducing inflammation throughout the body. These oils are thyme, clove, cinnamon, fennel, eucalyptus, bergamot, and rose.

In addition, many essential oils can improve blood flow and circulation by reducing the amount of nitric oxide in the body. Nitric oxide has been linked to the type of inflammation attributed to high blood pressure, heart disease, and diabetes.

Weight Loss

Essential oils are effective at promoting weight loss. They do this by suppressing appetite, boosting metabolism and increasing the rate at which the body burns fat, curbing sugar and carb cravings, reducing bloating and water retention, and general detoxification. By regulating emotions, they can also reduce overeating linked to stress and other emotional issues.

Multi Purpose

The properties of essential oils are complex and work together in a synergistic fashion on multiple systems. This is why just one essential oil can be used for so many different symptoms. When used in combination, their power increases exponentially.

Quality of Essential Oils

The quality of essential oils is highly variable and as a consumer it can be difficult to determine if you are purchasing a high-quality oil. Essential oils are regulated by the Food and Drug Administration (FDA) and they are classified either as cosmetics, food, or drugs, depending upon their intended use. How the FDA classifies them varies, but most are not considered drugs by the FDA and are instead considered cosmetics or fragrances.

Many factors go into the production of essential oils and things like soil quality, growing conditions, and harvesting methods can all make a difference in the final product. In addition, some companies selling essential oils will dilute or adulterate the oil in some way and this can be hard to identify. It is important to find pure, unadulterated oils because otherwise the oil will not be as potent and may not have the intended effects.

The grade of the essential oil is another factor to consider. While essential oils are not actually "graded" by an outside regulatory body, many essential oils are sold in different grades and these can be a good indicator of quality. The higher graded oils are typically purer, single-sourced, and higher quality and, of course, command higher prices.

Buying Essential Oils

Once you enter into the world of aromatherapy, you'll soon realize that along with the many enticing scents, there is also a multitude of purchasing sources. It can be confusing to decide where to purchase your essential oils. The following are some issues to consider.

Make sure you read the labels and the product descriptions. Know what you are purchasing, whether online or offline. Look for labels containing the words "pure" or "100 percent" essential oil. Anything else, and the product is most likely diluted with either a carrier oil or other additives, and you will not be getting a true essential oil. Also, some companies may be trying to pass off synthetic oils as true essential oils.

In addition, know your prices before you shop. By knowing which oils you would expect to pay more for, such as those considered rare or exotic, you'll have a better idea whether or not the asking price is accurate. The price of aromatherapy products and ingredients will vary but still should fall within some commonly accepted price ranges.

Steer clear from essential oils that are bottled using clear or plastic bottles. Light and plastic can damage and contaminate these oils. Dark-colored glass bottles—amber or dark blue—are best for storage.

Also, watch out for dust. Why? An essential oil's properties, and therefore its effectiveness, will start to diminish over time. Dust is a sign that the bottle's already spent some time on the shelf and perhaps isn't as fresh as it could be. Of course, if you're purchasing online you won't be able to see this factor.

Once you have found a source for essential oils that you're comfortable with, it's wise to become a loyal customer. Since essential oils are all-natural, consistency can be a challenge. You might find a single source offers you the best chance of getting a consistent high-quality product.

Check out the resources section at the back of the book for a list of sources for high-quality essential oils.

Storing Your Oils

How you store your oils will have a big impact on their shelf life. Left under the wrong conditions, these delicate oils can break down and become worthless.

First, be sure your oils are kept in dark-colored glass bottles. This will protect them from harmful ultraviolet light. This is most likely how you purchased them, but if for some reason they were purchased in clear or plastic bottles then be sure to transfer them as soon as possible.

Next, remember that essential oils do not like extremes of heat or light. Do no place them out on a window shelf in the sun, regardless of how pretty they may look decorating your window! This will cause them to deteriorate very quickly. So find a cool dark place and keep them there. Some people keep their essential oils in the refrigerator, especially citrus oils which like it cool. Just make sure they don't get confused with your edible items. Be sure to take them out of the refrigerator a couple of hours before using them to allow any fatty particles to re-dissolve. Give them a shake before using to make sure all of the particles are dissolved evenly.

Remember to place the caps back tightly on the bottles when you have finished with them. Otherwise, the essential oils will evaporate.

Proper storage of essential oils is also a significant concern for safety reasons. Some essential oils are toxins if ingested and they are also flammable. The smells may be very enticing to someone, such as a child, who does not know what they are and they may attempt to eat them. This can have disastrous results. In terms of flammability, the distillation process makes it very similar to alcohol, and highly sensitive to open flames. For these reasons, be sure to store the oils out of reach of children.

Safety Considerations

Although essential oils are natural products that have been used for thousands of years, there are certain safety precautions to keep in mind.

First, the majority of essential oils are meant to be inhaled or applied topically. They are not meant to be ingested and can be toxic or even fatal if consumed.

Many essential oils can irritate the skin and mucous membranes. Pure undiluted oil should not be used directly on the skin. They are extremely powerful in this form and may cause an adverse reaction.

As a general rule, essential oil therapy should not be used in pregnancy, by lactating women, or young children. There are some exceptions to this rule but it is a good idea to check with your health practitioner before using any essential oil in these situations.

Keep all essential oils out of reach of children. Remember to treat essential oils as the medicines that they are. While they may smell good, they are extremely powerful and are never safe for children to ingest.

If you are suffering from any chronic health condition or take prescription medications then please check with your health practitioner before embarking on essential oil therapy to avoid possible contraindications. People with asthma, epilepsy, kidney or liver conditions should be especially careful when using essential oils.

Patch Test

Some essential oils can cause an allergic reaction or sensitization in some individuals.

Before using any essential oil it is important to do a patch test on a small area of skin to check for any type of allergic or other type of adverse reaction. Apply a small amount of diluted oil to your inner arm. Wait 24 hours to check for redness or irritation to develop. If it does, wash the area with soap and water and do not use the oil.

Phototoxicity

Sun sensitivity, otherwise known as phototoxicity, can occur when using certain essential oils. What this means is that your skin will be more sensitive to sunlight and more prone to sunburn in the period following use of the oil. The following oils have been known to induce phototoxicity and should not be applied to skin that will be in direct sunlight within 24 hours: angelica, bergamot, citrus oils, cumin, dill, tagetes, yuzu.

Ingestion of Essential Oils

As stated previously, it is not recommended that essential oils should ever be ingested. Do so only under the guidance of a qualified professional.

In case of accidental ingestion, call your local poison control center.

Hazardous Oils

Certain essential oils should not be used in home aromatherapy preparations. These include wormwood, pennyroyal, camphor, wintergreen, and bitter almond. Use of these essential oils should only be used with the guidance of a qualified professional.

Aroma Types

Below is a helpful chart which organizes some common essential oils by aroma types.

Aroma Category
Camphorous/Medicinal: Cajuput, Eucalyptus, Tea Tree
Citrus: Lemon, Lime, Orange, Tangerine
Earthy: Oakmoss, Vetiver, Patchouli
Floral: Jasmine, Lavender, Neroli, Rose
Herbaceous: Basil, Marjoram, Rosemary
Minty: Peppermint, Spearmint
Oriental: Ginger, Patchouli
Spicy: Clove, Cinnamon, Nutmeg
Woodsy: Cedar, Pine

In general, essential oils in the same category blend well with one another. In addition, here are a few general guidelines for what works well together:

- Florals blend well with citrusy, spicy, and woodsy scents.
- Minty scents blend well with citrus, earthy, herbaceous, and woodsy scents.
- Oriental scents blend well with citrus, florals, and spicy scents.
- Spicy scents blend well with citrus, florals, and orientals.
- Woodsy oils blend well with most other scents.

Part II

Recipes

Instructions for recipes: Fill cool-air diffuser with water according to manufacturer's instructions. Place essential oil drops in diffuser. Turn on diffuser and diffuse oils for desired amount of time.

Energize

Best essential oils for energy are: basil, citrus, cedarwood, eucalyptus, ginger, grapefruit, lemon, orange, peppermint, pine, rosemary, vanilla

When you're feeling fatigued and need an energetic pick-me-up, try one of these blends instead of a cup of coffee.

Quick Pick-Me-Up

Blend:

    5 drops peppermint

    3 drops frankincense

    3 drops grapefruit

Energy Booster

Blend:

    4 drops peppermint

    4 drops rosemary

    2 drops ginger

    2 drops basil

Get the Power

Blend:

    6 drops grapefruit

    3 drops basil

    3 drops bergamot

    2 drops lavender

High-Voltage Blend

Blend:

    3 drops lemon

    3 drops grapefruit

    2 drops spearmint

Energizer Bunny

Blend:

    3 drops orange

    2 drops thyme

    2 drops basil

Energy Shot

Blend:

    3 drops cypress

    2 drops bergamot

    2 drops orange

Stimulating Blend

Blend:

    3 drops peppermint

    3 drops lemon

    3 drops grapefruit

Wake Me Up

Blend:

    3 drops black pepper

- 2 drops thyme
- 2 drops rosemary
- 1 drop lemon

Feeling Perky

Blend:

- 4 drops lemongrass
- 3 drops eucalyptus
- 1 drop rosemary

Invigorating

Blend:

- 3 drops orange
- 2 drops coriander
- 2 drops clary sage
- 1 drop jasmine
- 1 drop vetiver

Sleep

Best essential oils for sleep are: angelica, basil, bergamot, cedarwood, lavender, mandarin orange, neroli, patchouli, roman chamomile, rose, sandalwood, sweet marjoram, valerian, vetiver, ylang ylang

Try these soothing blends to help send you into dreamland.

Dream Time

Blend:

    3 drops cedarwood

    2 drops ylang ylang

    2 drops sandalwood

Sweet Dreams

Blend:

    3 drops lavender

    3 drops valerian

    2 drops sweet marjoram

    2 drops rose

Good Night

Blend:

    3 drops lavender

    3 drops cedarwood

    2 drops mandarin orange

No Need to Count Sheep

Blend:

    3 drops Roman chamomile

    3 drops sandalwood

2 drops marjoram

Land of Nod
Blend:

    3 drops vetiver

    3 drops rose

    2 drops lavender

Drift Away
Blend:

    3 drops Roman chamomile

    3 drops frankincense

    2 drops bergamot

Lullaby
Blend:

    3 drops valerian

    3 drops wild orange

    2 drop bergamot

Cloud Nine
Blend:

    3 drops lavender

    3 drops chamomile

    2 drops vetiver

Rest for the Wicked

Blend:

    3 drops patchouli

    3 drops sandalwood

    2 drops wild orange

Relaxation and Calmness

Best essential oils to promote relaxation and calmness are: clary sage, helichrysum, mandarin orange, marjoram, neroli, Roman chamomile, rose, spikenard, vetiver, ylang ylang

These blends are perfect for when you want to slow down and relax. Try these blends to unwind at the end of a long day.

Unwind

Blend

- 3 drops marjoram
- 3 drops rose
- 2 drops vetiver

Destress Blend

Blend

- 3 drops clary sage
- 3 drops ylang ylang
- 2 drops marjoram

Stress Melter

Blend

- 3 drops Roman chamomile
- 2 drops jasmine
- 2 drops rose

Calm the H@ll Down

Blend

    3 drops mandarin orange

    3 drops neroli

    2 drops cedarwood

Tension Melter

Blend

    3 drops clary sage

    3 drops bergamot

    2 drops Roman chamomile

Peacefulness

Blend

    3 drops marjoram

    3 drops lavender

    2 drops vetiver

Serenity Now

Blend

    3 drops ylang ylang

    3 drops Mandarin orange

    2 drops cedarwood

Peace and Quiet

Blend

- 3 drops helichrysum
- 3 drops rose
- 2 drops sandalwood

Quiet Moments

Blend

- 3 drops neroli
- 3 drops ylang ylang
- 2 drops lavender

Take It Easy

Blend

- 3 drops cedarwood
- 3 drops ylang ylang
- 2 drops orange

Fear No More

Blend

- 4 drops clary sage
- 4 drops Roman chamomile
- 3 drops vetiver

Mood Enhancement

Best essential oils for mood enhancement are: bergamot, geranium, grapefruit, jasmine, lavender, lemon, peppermint, rose, sweet orange, sandalwood, scotch pine, vetiver, vanilla, ylang ylang

Try these blends for a quick lift when you are feeling down and to promote overall well-being.

Lift Me Up

Blend

    3 drops jasmine

    3 drops basil

    2 drops bergamot

Out of the Dumps

Blend

    3 drops clary sage

    3 drops ylang ylang

    2 drops sweet orang

Feeling Groovy

Blend

    3 drops spearmint

    3 drops vanilla

    2 drops sweet orange

Brighten Up

Blend

- 3 drops grapefruit
- 3 drops rose
- 2 drops bergamot

Feeling Happy

Blend

- 3 drops rosemary
- 3 drops geranium
- 2 drops sweet orange

It's a Beautiful Day

Blend

- 3 drops rose
- 3 drops sweet orange
- 2 drops patchouli

Joy to the World

Blend

- 3 drops bergamot
- 3 drops frankincense
- 2 drops orange

Positivity

Blend

      3 drops clary sage

      3 drops ylang ylang

      2 drops lavender

On Top of the World

Blend

      3 drops lavender

      3 drops orange

      2 drops ylang ylang

Confidence Booster

Blend

      4 drops orange

      4 drops grapefruit

      2 drops jasmine

Sunny Side Up

Blend

      3 drops jasmine

      3 drops lemon

      2 drops chamomile

Anger Release

Blend

- 4 drops lavender
- 4 drops geranium
- 3 drops clary sage

Sadness Lifter

Blend

- 3 drops rose
- 3 drops cypress
- 3 drops helichrysum
- 2 drops frankincense

Happiness Blend

Blend

- 4 drops orange
- 4 drops grapefruit
- 3 drops neroli

Stress Relief

Best essential oils for stress relief are: bergamot, chamomile, cinnamon, cedarwood, clary sage, fennel, frankincense, geranium, lavender, lemon, lemongrass, marjoram, orange, rose, rosewood, sandalwood, vetiver, ylang ylang

These blends will help calm the nerves and soothe away a stressful day.

Peace and Tranquility

Blend

>3 drops clary sage
>
>3 drops orange
>
>2 drops frankincense

Serenity Blend

Blend

>3 drops lavender
>
>3 drops ylang ylang
>
>2 drops chamomile

Stress Melter

Blend

>3 drops bergamot
>
>3 drops ylang ylang
>
>2 drops lemon

Stress Away Blend

Blend

>3 drops sweet orange
>
>3 drops patchouli
>
>2 drops grapefruit

Stress Less

Blend

>3 drops clary sage
>
>2 drops lavender
>
>2 drops lemon

Mellow Out

Blend

>3 drops vetiver
>
>3 drops chamomile
>
>2 drops rose

Relax and Decompress

Blend

>3 drops bergamot
>
>2 drops geranium
>
>2 drops frankincense

Ultimate Unwind Blend

Blend

>3 drops lavender
>
>3 drops rose
>
>2 drops chamomile

De-stressful Situation

Blend

> 3 drops rosewood
>
> 2 drops cinnamon
>
> 2 drops marjoram
>
> 1 drop orange

Time to Chill

Blend

> 3 drops jasmine
>
> 3 drops grapefruit
>
> 2 drops vetiver

Calming Blend

Blend

> 5 drops ylang ylang
>
> 5 drops patchouli
>
> 3 drops orange
>
> 3 drops bergamot

Colds and Congestion

Best essential oils for colds and congestion are: eucalyptus, frankincense, lemon, oregano, peppermint, rosemary

These blends will provide relief from the symptoms of the common cold and nasal congestion.

Sinus Relief Blend

Blend

- 2 drops peppermint
- 2 drops eucalyptus
- 2 drops oregano
- 2 drops rosemary

Feel Better Blend

Blend

- 3 drops eucalyptus
- 3 drops peppermint
- 2 drops jasmine

Decongest Blend

Blend

- 3 drops cedarwood
- 3 drops eucalyptus
- 2 drops lavender

Breathe Easy Blend

Blend

- 3 drops tea tree oil
- 3 drops eucalyptus

2 drops bergamot

Cure for the Common Cold
Blend

    3 drops lemon

    3 drops tea tree

    3 drops lavender

Cold Away Blend
Blend

    3 drops lavender

    3 drops rosemary

    2 drops orange

Congestion Relief
Blend

    3 drops peppermint

    3 drops eucalyptus

    2 drops lavender

Stuffy Nose Fighter
Blend

    3 drops sandalwood

    3 drops eucalyptus

    2 drops lemongrass

top a Cold Blend

lend

    3 drops basil

    3 drops eucalyptus

    2 drops peppermint

Immune System Boosters

Best essential oils to boost immune system are: bergamot, cinnamon, eucalyptus, frankincense, grapefruit, lavender, lemon, oregano, peppermint, pine, rosemary, sage, tea tree, thyme, valor

These blends promote a healthy and strong immune system.

Get Healthy Blend

Blend

    3 drops cinnamon

    2 drops lemon

    2 drops eucalyptus

    2 drops rosemary

Immunity Blend

Blend

    3 drops oregano

    3 drops lemon

    2 drops thyme

Strong and Healthy

Blend

    3 drops sage

    2 drops eucalyptus

    2 drops lavender

Resistance Blend

Blend

    2 drops clove

  2 drops rosemary

  2 drops orange

  2 drops cinnamon

Feeling Healthy Blend

Blend

  3 drops pine

  2 drops cardamom

  2 drops cassia

Robust Blend

Blend

  3 drops grapefruit

  2 drops lemon

  2 drops tea tree

  1 drop frankincense

Fresh and Active Blend

Blend

  3 drops lavender

  2 drops geranium

  2 drops rosemary

Sickness No More

Blend

    3 drops sage

    3 drops bergamot

    2 drops valor

Immune Support Blend

Blend

    3 drops rosemary

    3 drops lemon

    2 drops cinnamon

Headache Relief

Best essential oils to relief headaches are: bergamot, chamomile, clary sage, eucalyptus, ginger, lavender, peppermint, rosemary

Fight headaches the natural way these blends.

Bye Bye Pain

Blend

- 3 drops peppermint
- 3 drops rosewood
- 2 drops lavender
- 2 drops rosemary

Migraine Relief Blend

Blend

- 3 drops peppermint
- 3 drops eucalyptus
- 2 drops rosemary
- 2 drops lavender

Goodbye Headache Blend

Blend

- 3 drops lavender
- 3 drops lemongrass
- 3 drops spearmint

Relax and Soothe Blend

Blend

- 3 drops frankincense
- 3 drops peppermint

>　　2 drops eucalyptus
>
>　　2 drops lavender

Cluster F@ck Blend

Blend

>　　3 drops rosemary
>
>　　3 drops chamomile
>
>　　2 drops lemongrass

Tension Headache Tamer

Blend

>　　3 drops eucalyptus
>
>　　3 drops clary sage
>
>　　2 drops lavender

Stress Headache Relief

Blend

>　　3 drops lavender
>
>　　3 drops chamomile
>
>　　2 drops clary sage

Headache Relief Blend

Blend

>　　3 drops rosemary
>
>　　3 drops bergamot
>
>　　2 drops ginger

Oh My Aching Head Blend

Blend

    3 drops jasmine

    3 drops rosemary

    2 drops peppermint

Better than Aspirin Blend

Blend

    3 drops lavender

    3 drops rose

    2 drops marjoram

    2 drops bergamot

Cognitive Function

Best essential oils to boost brain power are: basil, bergamot, eucalyptus, frankincense, juniper, lemon, peppermint, pine, rosemary, sage, vetiver

These blends will help you focus and feel sharp.

Put on Your Thinking Cap Blend

Blend

> 3 drops rosemary
>
> 3 drops clary sage
>
> 3 drops juniper

Brain Power Blend

Blend

> 3 drops basil
>
> 3 drops rosemary
>
> 2 drops cypress

They Call Me Einstein Blend

Blend

> 3 drops eucalyptus
>
> 3 drops peppermint
>
> 3 drops basil

Just Focus Blend

Blend

> 3 drops peppermint
>
> 3 drops vetiver
>
> 2 drops lemon

Cognitive Power Blend

Blend

- 3 drops rosemary
- 3 drops sage
- 3 drops lavender

Think, Think, Think!

Blend

- 3 drops rosemary
- 3 drops thyme
- 3 drops rose

Concentration Blend

Blend

- 3 drops peppermint
- 3 drops basil
- 3 drops helichrysum

Sharp as a Tack Blend

Blend

- 2 drops peppermint
- 2 drops rosemary
- 2 drops lemon
- 2 drops basil
- 2 drops frankincense

Brainiac Blend

Blend

    3 drops basil

    3 drops frankincense

    3 drops vetiver

Insect Repellant

The best essential oils to repel insects such as mosquitoes, flies, moths, fruit flies are: clove, cedarwood, citronella, eucalyptus, geranium, lemongrass, peppermint, rosemary, tea tree

Keep away those pests ruining your outdoor experience with these blends.

Go Away Bugs Blend

Blend

4 drops geranium

3 drops citronella

2 drops eucalyptus

2 drops lemon

Insect Repellant Blend

Blend

4 drops citronella

3 drops rosemary

3 drops lavender

Buzz Off Blend

Blend

3 drops spearmint

3 drops lemongrass

2 drops citronella

2 drops basil

Bug Off Blend

Blend

3 drops eucalyptus

> 3 drops tea tree
>
> 2 drops orange

Fly Fly Away Blend

Blend

> 3 drops cedarwood
>
> 3 drops lemongrass
>
> 2 drops pine
>
> 2 drops lemon

Bugger Off Blend

Blend

> 3 drops rose
>
> 2 drops geranium
>
> 2 drops juniper
>
> 2 drops thyme

Just Get Away Blend

Blend

> 3 drops citronella
>
> 3 drops peppermint
>
> 3 drops lemongrass
>
> 2 drops tea tree

Meditation

Best essential oils for meditation are: cedarwood, chamomile, clary sage, frankincense, helichrysum, lavender, myrrh, patchouli, rose, sage, sandalwood, vetiver

Calm the mind and spirit with these blends.

Just Say Ohm

Blend

    3 drops sandalwood

    3 drops frankincense

    3 drops bergamot

Serenity Now

Blend

    3 drops patchouli

    3 drops rose

    3 drops clary sage

Contemplation Blend

Blend

    4 drops frankincense

    4 drops helichrysum

    2 drops lavender

Reflection Blend

Blend

    3 drops clary sage

    3 drops cedarwood

    2 drops neroli

Clear Your Mind

Blend

>3 drops lavender
>
>3 drops bergamot
>
>2 drops vetiver

Meditation Blend

Blend

>3 drops cedarwood
>
>2 drops tangerine
>
>2 drops patchouli
>
>2 drops bergamot
>
>2 drops ylang ylang

Peaceful Feeling Blend

Blend

>3 drops frankincense
>
>3 drops sweet orange
>
>2 drops sage

Everything Zen

Blend

>4 drops sandalwood
>
>2 drops cedarwood
>
>2 drops lemon

Appetite Suppressant/Fight Hunger Cravings

Best essential oils for weight loss are: bergamot, cinnamon, eucalyptus, frankincense, ginger, grapefruit, jasmine, lavender, lemon, orange, peppermint, rosemary, sandalwood

Fight hunger and support weight loss efforts with these blends.

Totally Satisfied Blend

Blend

- 3 drops peppermint
- 3 drops lemon
- 2 drops rosemary

Bye Bye Hunger Blend

Blend

- 4 drops grapefruit
- 3 drops rosemary
- 3 drops tangerine

Not Even Hungry Blend

Blend

- 3 drops rosemary
- 3 drops ginger
- 2 drops bergamot

Appetite Suppressant Blend

Blend

    3 drops cypress

    3 drops lavender

    2 drops orange

    2 drops sandalwood

Don't Want Any Blend

Blend

    3 drops eucalyptus

    3 drops lemon

    2 drops cypress

Diet Support Blend

Blend

    3 drops fennel

    3 drops rosemary

    2 drops spearmint

    2 drops ginger

Air Fresheners

Best essential oils to naturally freshen the air are: bergamot, cinnamon, jasmine, lavender, lemon, orange, rose, rosemary

These natural essential oil blends will freshen indoor air without the chemical toxins of commercial air fresheners.

Light and Breezy Blend

Blend

- 5 drops peppermint
- 5 drops wild orange
- 3 drops bergamot

Smells Like Heaven

Blend

- 4 drops wild orange
- 4 drops lemon
- 3 drops grapefruit

Fresh Fields

Blend

- 4 drops lavender
- 3 drops lemon
- 3 drops rosemary

Breath of Fresh Air

Blend

- 3 drops eucalyptus
- 3 drops cedarwood
- 3 drops lemon

Captain Fresh Blend

Blend

>4 drops white fir
>
>4 drops cedarwood
>
>3 drops frankincense

You Must Be Heaven Scent

Blend

>3 drops jasmine
>
>3 drops grapefruit
>
>3 drops wild orange

Freshness in a Bottle

Blend

>4 drops lime
>
>3 drops bergamot
>
>3 drops ylang ylang
>
>2 drops rose

Clear the Air Blend

Blend

>4 drops lavender
>
>4 drops clary sage
>
>3 drops lemon

Fresh Aroma

Blend

- 4 drops jasmine
- 4 drops chamomile
- 3 drops lemon
- 2 drops spearmint

Beat the Heat

Best essential oils to help beat the heat are: eucalyptus, grapefruit, lemon, peppermint, spearmint

When the weather heats up diffuse these cooling blends.

Cool Off Blend

Blend

    3 drops peppermint

    3 drops lavender

    2 drops rose

Light and Airy Blend

Blend

    3 drops eucalyptus

    3 drops lemon

    2 drops lavender

Things Are Getting Steamy Blend

Blend

    4 drops spearmint

    4 drops lemongrass

    2 drops wild orange

Beat the Heat Blend

Blend

    4 drops spearmint

    3 drops jasmine

    2 drops peppermint

Cool Mist Blend

- 3 drops grapefruit
- 3 drops rose
- 2 drops peppermint

Positive Thinking

Best essential oils for staying in a positive state of mind are: bergamot, clary sage, coriander, cypress, eucalyptus, frankincense, ginger, hyssop, jasmine, lemon, melissa, myrrh, orange, patchouli, pine, rosemary, sweet marjoram, tangerine, tea tree, vetiver, ylang-ylang

Diffuse these blends to get in a more positive state of mind.

Just Keep Swimming Blend

Blend

>3 drops cypress
>
>3 drops frankincense
>
>2 drops tangerine

You Are My Sunshine Blend

Blend

>3 drops orange
>
>3 drops ginger
>
>3 drops vetiver

Optimism Blend

Blend

>3 drops bergamot
>
>3 drops orange
>
>2 drops eucalyptus

Think Positive Blend

Blend

>3 drops peppermint
>
>3 drops rosemary

2 drops lime

2 drops grapefruit

2 drops lemon

Positively Winning Blend

3 drops basil

3 drops rosemary

3 drops spearmint

2 drops ginger

Things Are Looking Up Blend

3 drops juniper

3 drops wild orange

2 drops lavender

2 drops sage

Always a Silver Lining Blend

3 drops cinnamon

3 drops orange

2 drops peppermint

Great Expectations Blend

Blend

    3 drops frankincense

    3 drops black pepper

    3 drops lime

Rosy Outlook Blend

Blend

    3 drops pine

    3 drops basil

    3 drops rose

    2 drops lavender

Anxiety

The best essential oils to relieve anxiety include: bergamot, cedarwood, cinnamon, chamomile, clary sage, eucalyptus, frankincense, grapefruit, lavender, lime, rose, vetiver, ylang ylang

Diffuse these blends to naturally relive anxiety.

Peaceful Blend

Blend

- 4 drops clary sage
- 4 drops lavender
- 2 drops lemon

Just Breath Blend

Blend

- 4 drops cinnamon
- 4 drops orange
- 2 drops rose

Ten Deep Breaths Blend

Blend

- 4 drops lime
- 2 drops mandarin
- 2 drops lavender

Anti-Anxiety Blend

Blend

- 5 drops eucalyptus
- 3 drops bergamot
- 3 drops chamomile

Anti-Angst Blend

Blend

>4 drops rose
>
>3 drops orange
>
>3 drops sandalwood

Cool and Collected Blend

Blend

>3 drops ylang ylang
>
>3 drops clary sage
>
>3 drops lavender

Put Myself at Ease Blend

Blend

>3 drops jasmine
>
>3 drops grapefruit
>
>2 drops tangerine
>
>2 drops ylang ylang

Tranquility Now Blend

Blend

>4 drops frankincense
>
>4 drops jasmine
>
>2 drops lemon
>
>2 drops orange

Joy and Security Blend

Blend

- 4 drops lemongrass
- 3 drops lavender
- 3 drops vetiver

Relax the Nerves

Blend

- 4 drops lavender
- 4 drops vetiver
- 2 drops orange
- 2 drops frankincense
- 1 drop copaiba

Insecurity Soother

Blend

- 4 drops cedarwood
- 4 drops bergamot
- 2 drops sandalwood

Panic Buster

Blend

- 5 drops lavender
- 3 drops valor
- 3 drops wild orange
- 2 drops copaiba
- 2 drops frankincense

Depression

The best essential oils to combat depression are: bergamot, basil, chamomile, clary sage, frankincense, geranium, grapefruit, jasmine, lavender, lemon, neroli, orange, patchouli, rose, sandalwood, valerian, ylang ylang

Try these blends to help your mood and ease depression.

Fight the Blues Blend

Blend

    3 drops sandalwood

    3 drops rose

    2 drops orange

Lift Me Up Blend

Blend

    6 drops bergamot

    3 drops clary sage

Out of the Clouds Blend

Blend

    3 drops grapefruit

    3 drops lavender

    3 drops ylang ylang

No More Blues Blend

Blend

    3 drops jasmine

    3 drops frankincense

    2 drops lemon

2 drops orange

Anti-depressant Blend

Blend

4 drops bergamot

4 drops clary sage

3 drops vetiver

Let There Be Hope Blend

Blend

4 drops neroli

4 drops ylang ylang

3 drops sandalwood

3 drops lavender

Don't Tell Me to Cheer Up Blend

Blend

4 drops rose

3 drops orange

2 drops bergamot

2 drops patchouli

Wish Upon a Star Blend

Blend

4 drops lavender

4 drops rose

2 drops geranium

2 drops valerian

Uplifting Blend
Blend

4 drops sandalwood

4 drops rose

2 drops lemon

2 drops wild orange

Indigestion and Nausea

The best essential oils to calm your stomach and ease nausea are: bergamot, black pepper, cardamom, coriander, fennel, ginger, grapefruit, lavender, lemon, Melissa, nutmeg, peppermint, Roman chamomile, spearmint

Use these blends to soothe queasiness, motion sickness, indigestion, heartburn, and nausea.

Upset Stomach Blend

Blend

- 4 drops bergamot
- 4 drops cardamom
- 3 drops grapefruit
- 3 drops spearmint
- 2 drops geranium

Morning Sickness Blend

Blend

- 4 drops ginger
- 3 drops grapefruit
- 3 drops lavender
- 2 drops Roman chamomile

Anti-Nausea Blend

Blend

- 4 drops spearmint
- 4 drops lavender
- 3 drops lemon

Calm the Queasiness Blend

Blend

    4 drops bergamot

    4 drops orange

    3 drops ginger

That Queasy Feeling Blend

Blend

    4 drops ginger

    4 drops lavender

    2 drops peppermint

    2 drops spearmint

    2 drops basil

Nervous Stomach Blend

Blend

    4 drops basil

    4 drops coriander

    2 drops peppermint

    2 drops jasmine

Motion Sickness Relief

Blend

    4 drops Roman chamomile

    4 drops ginger

    4 drops spearmint

Holiday/Winter Blends

The best essential oils to spread holiday cheer are: cedar, cinnamon, clove, Douglas fir, frankincense, myrrh, orange, pine, white fir

Diffuse these blends to get in the holiday spirit.

Holly Jolly Blend

Blend

   5 drops cinnamon

   4 drops white fir

   4 drops wild orange

Merry and Bright Blend

Blend

   4 drops frankincense

   4 drops myrrh

   4 drops wild orange

Silent Night Blend

Blend

   5 drops wild orange

   4 drops clove

   4 drops cinnamon

   2 drops frankincense

Ho Ho Ho Blend

Blend

    5 drops Douglas fir

    5 drops cedar

    2 drops nutmeg

Winter Wonderland Blend

Blend

    5 drops eucalyptus

    5 drops juniper

    3 drops sage

Most Wonderful Time of the Year Blend

Blend

    5 drops peppermint

    4 drops ylang ylang

    3 drops sage

Happy New Year Blend

Blend

    4 drops clove

    4 drops cinnamon

    3 drops cardamom

    3 drops wild orange

Oh Christmas Tree Blend

5 drops white fir

5 drops bergamot

3 drops cassia

3 drops ginger

Let It Snow Blend

3 drops cinnamon

3 drops orange

3 drops ginger

Summertime Blends

The best essential oils for summer are: basil, bergamot, chamomile, citronella, clary sage, coriander, cypress, geranium, helichrysum, hyssop, jasmine, lavender, lemon, lemongrass, lime, mandarin, Melissa, neroli, orange, Roman chamomile, rose, thyme, ylang ylang

Celebrate summer with these sunny blends.

Here Comes the Sun

Blend

    8 drops mandarin

    4 drops ginger

    4 drops patchouli

Sunkissed Blend

Blend

    8 drops yuzu

    4 drops rose

    4 drops jasmine

    2 drops Roman chamomile

Day at the Beach Blend

Blend

    4 drops grapefruit

    4 drops lavender

    3 drops lemon

    3 drops spearmint

You Are My Sunshine Blend

Blend

4 drops cypress

4 drops lemon

3 drops lavender

3 drops juniper

Sunny Day Blend

Blend

4 drops lemongrass

4 drops wild orange

2 drops rose

2 drops spearmint

Ocean Breeze Blend

Blend

4 drops lime

4 drops tangerine

3 drops spearmint

Beach Life Blend

Blend

3 drops bergamot

3 drops eucalyptus

2 drops lavender

2 drops rosemary

Waves on the Beach Blend

Blend

       3 drops grapefruit

       3 drops juniper

       3 drops orange

       1 drop peppermint

Sea You Soon Blend

Blend

       4 drops lemon

       4 drops lime

       3 drops basil

       2 drops jasmine

       1 drop spearmint

Autumn Blends

These are some of the best essential oils for the fall season: cassia, cedarwood, cinnamon, clove, eucalyptus, frankincense, ginger, juniper berry, nutmeg, orange, patchouli, rosemary, sage, tangerine, white fir

These blends will bring to mind apples, pumpkins, and crisp fall leaves.

Falling Leaves Blend

Blend

- 4 drops cassia
- 3 drops wild orange
- 3 drops white fir

Pumpkin Spice Blend

Blend

- 4 drops cinnamon
- 3 drops nutmeg
- 3 drops clove

Apple Pie Blend

Blend

- 3 drops cassia
- 3 drops cinnamon
- 2 drops nutmeg
- 2 drops ginger

Fall Breeze

Blend

    4 drops juniper berry

    4 drops eucalyptus

    2 drops sage

Leaves Falling Down

Blend

    4 drops cardamom

    2 drops cinnamon

    2 drops clove

    2 drops wild orange

Hello Fall Blend

Blend

    4 drops wild orange

    3 drops frankincense

    3 drops clove

    2 drops ginger

Harvest Time Blend

Blend

    3 drops frankincense

    3 drops white fir

    2 drops cedarwood

October Sky Blend
Blend

    4 drops cypress

    3 drops white fir

    3 drops sandalwood

    2 drops cinnamon

Fall Is in the Air
Blend

    3 drops bergamot

    2 drops lemon

    2 drops orange

    2 drops white fir

Springtime Blends

The best essential oils to celebrate spring are: bergamot, eucalyptus, frankincense, grapefruit, geranium, jasmine, lavender, lemon, lemongrass, lime, orange, rose, spearmint, tangerine, ylang ylang

Spring is in the air with these fresh blends.

Hello Spring Blend

Blend

    3 drops lavender

    3 drops orange

    3 drops geranium

Spring Has Sprung Blend

Blend

    3 drops lemon

    3 drops rosemary

    3 drops lavender

New Beginnings Blend

Blend

    3 drops jasmine

    3 drops lime

    2 drops geranium

Flowers, Flowers, Everywhere Blend

Blend

    4 drops ylang ylang

    3 drops rose

    2 drops lavender

2 drops orange

First Buds Blend

Blend

3 drops clary sage

3 drops jasmine

2 drops geranium

Springtime Blend

Blend

3 drops lavender

3 drops geranium

2 drops ylang ylang

Bloom Blend

Blend

4 drops rose

3 drops lemon

3 drops frankincense

1 drop ylang ylang

Spring Is in the Air Blend

Blend

3 drops lavender

3 drops lemon

2 drops peppermint

Resources and References

Books on Essential Oils and Aromatherapy

For more information on aromatherapy and essential oils check out these sources of information.

- *Essential Oils for Beginners* by Julia Grady
- *375 Essential Oils and Hydrosols* by Jeanne Rose
- *Advanced Aromatherapy* by Kurt Schnaubelt
- *Aromatherapy for the Healthy Child* by Valerie Ann Wormwood
- *The Art of Aromatherapy* by Robert Tisserand
- *The Complete Book Essential Oils & Aromatherapy* by Valerie Ann Worwood
- *Essential Oil Safety: A Guide for Health Care Practitioners* by Robert Tisserand and Tony Balacs
- *The Essential Oils Book* by Colleen K. Dodt
- *Holistic Aromatherapy for Animals* by Kristen Leigh Bell
- *Hydrosols: The Next Aromatherapy* by Suzanne Catty
- *The Illustrated Encyclopedia of Essential Oils* by Julia Lawless
- *Practical Aromatherapy: How to Use Essential Oils to Restore Vitality* by Shirley Price

Sources of Essential Oils

Here are a list of recommended sources for purchasing essential oils and other supplies.

Arlys Naturals

All natural essential oils and skin care products.

www.arlysnaturals.com

1-877-502-7597

Artisan Aromatics

Tests each essential oil for quality assurance.

http://ArtisanEssentialOils.com

1-828-835-2231

Liberty Natural Products

Grower, importer, and wholesaler.

www.libertynatural.com

1-800-289-8427

Mountain Rose Herbs

Established in 1987.

www.mountainroseherbs.com

1-800-879-3337